U. S. DEPARTMENT OF AGRICULTURE.

BUREAU OF PLANT INDUSTRY—BULLETIN NO. 195.

B. T. GALLOWAY, *Chief of Bureau.*

THE PRODUCTION OF VOLATILE OILS AND PERFUMERY PLANTS IN THE UNITED STATES.

BY

FRANK RABAK,

CHEMICAL BIOLOGIST, DRUG-PLANT, POISONOUS-PLANT, PHYSIO-
LOGICAL, AND FERMENTATION INVESTIGATIONS.

I0118985

British Library Cataloguing-in-Publication Data
A catalogue record for this book is available from the
British Library

Essential Oils

Essential oils are also known as volatile oils, ethereal oils, aetherolea, or simply as the 'oil of' the plant from which they are extracted, such as the oil of clove. An oil is 'essential' in the sense that it contains the characteristic fragrance of the plant that it is taken from. Essential oils do not form a distinctive category for any medicinal, pharmacological, or culinary purpose - and they are not essential for health, although they have been used medicinally in history. Although some are suspicious or dismissive towards the use of essential oils in healthcare or pharmacology, essential oils retain considerable popular use, partly in fringe medicine and partly in popular remedies. Therefore it is difficult to obtain reliable references concerning their pharmacological merits.

Medicinal applications proposed by those who sell or use medical oils range from skin treatments to remedies from cancer - and are generally based on historical efficacy. Having said this, some essential oils such as those of juniper and agathosma are valued for their diuretic effects. Other oils, such as clove oil or eugenol were popular for many hundreds of years in dentistry and as antiseptics and local anaesthetics. However as the use of essential oils has declined in evidence based medicine, older text-books are frequently our only sources for information! Modern works are less inclined to generalise; rather than referring to 'essential oils' as a class at all, they prefer to discuss specific compounds, such as methyl salicylate, rather than 'oil of wintergreen.'

Nevertheless, interest in essential oils has considerably revived in recent decades, with the popularity of aromatherapy, alternative health stores and massage. Generally, the oils are volatized or diluted with a carrier oil to be used in massage, or diffused in the air by a nebulizer, heated over a candle flame, or burned as incense. Their usage goes way back, and the earliest recorded mention of such methods used to produce essential oils was made by Ibn al-Baitar (1188-1248), an Andalusian physician, pharmacist and chemist. Different oils were claimed to have differing properties; some to have an uplifting and energizing effect on the mind such as grapefruit and jasmine, whilst others such as rose lavender have a reputation as de-stressing and relaxing - and also, usefully, as an insect repellent.

The oils themselves are usually extracted by 'distillation', often by using steam -but some other processes include 'expression' or 'solvent extraction'. Distillation involves raw plant material (be that flowers, leaves, wood, bark, roots, seeds or peel) put into an alembic (distillation apparatus) over water. As the water is heated, the steam passes through the plant material, vaporizing the volatile compounds. The vapours flow through a coil, where they condense back to liquid, which is then collected in the receiving vessel. 'Expression' differs in that it usually merely uses a mechanical or cold press to extract the oil. Most citrus peel oils are made in this way, and due to the relatively large quantities of oil in citrus peel and low cost to grow and harvest the raw materials, citrus-fruit oils are cheaper than most other essential oils. 'Solvent extraction' is perhaps the most difficult of the three methods, and is generally used for flowers, which contain too little volatile oil to undergo expression. Instead, a solvent such as hexane or supercritical carbon dioxide is used to extract the oils.

These techniques have allowed essential oils to be used in all manner of products; from perfumes to cosmetics, soaps - and as flavourings for food and drinks as well as adding scent to incense and household cleaning products. The science, history and folkloric tradition of essential oils is incredibly fascinating - and a still much debated area. We hope the reader is inspired by this book to find out more.

LETTER OF TRANSMITTAL.

U. S. Department of Agriculture,
Bureau of Plant Industry,
Office of the Chief,
Washington, D. C., August 26, 1910.

Sir: I have the honor to transmit herewith and to recommend for publication as Bulletin No. 195 of the special series of this Bureau a manuscript by Mr. Frank Rabak, Chemical Biologist, entitled "The Production of Volatile Oils and Perfumery Plants in the United States," submitted by Dr. R. H. True, Physiologist in Charge of the Office of Drug-Plant, Poisonous-Plant, Physiological, and Fermentation Investigations.

There is a steady demand for information concerning plants yielding materials used in the manufacture of perfumery products; also concerning the processes and apparatus required to utilize these oil-bearing plants. This line of agricultural work has not yet reached any marked development outside of the peppermint industry in Michigan, New York, and Indiana, but the outlook for a further growth of this branch of special agriculture seems worth consideration. Much experimental work will be required to determine the most favorable locations for operation, and practical experience in handling the crops and the special apparatus needed in utilizing them must be accumulated. However, the economic significance of this class of products seems likely to justify the efforts required.

Respectfully,

G. H. Powell,
Acting Chief of Bureau.

Hon. James Wilson,
Secretary of Agriculture.

CONTENTS.

195

ILLUSTRATIONS.

195
6

B. P. I.—603.

THE PRODUCTION OF VOLATILE OILS AND PERFUMERY PLANTS IN THE UNITED STATES.

INTRODUCTION.

The use of aromatics and perfumery dates back to the early ages when spices, balsams, asafetida, and other resinous exudations, many of which possess agreeable odors, were used for the purpose of scenting. The peculiar, agreeable aromas emanating from plants growing in their native habitats may be supposed to have early aroused the attention and admiration of the primitive peoples, although it may not have been known in what forms plants and flowers possessed their aromas. Before the art of distillation was known, the ancient peoples used the odoriferous plants and spices in their dried forms for their agreeable odors. Gradually, however, the development of special utensils for other domestic purposes may have resulted in the discovery of methods for the separation of odors from plants and plant products.

The use of distilling apparatus by the ancients in their endeavor to solve the problem of the transmutation of the elements and in other researches requiring the separation of volatile from nonvolatile substances antedates its use for the production of essential oils and perfumes, but it was probably learned at an early date that the odors present in plants and plant exudations were capable of separation because of their greater volatility when compared with the other constituents present. The first mention in ancient Greek writings of the separation of an odor from a crude substance is that of the oil of cedar, which was separated from the oleoresin by means of the crudest form of apparatus. This consisted of an open earthen kettle in which the oleoresin was boiled with water, the vapors of steam and oil being collected in layers of wool so placed that the steam from the kettle passed through the wool, which served as a condenser and retained the oil and water. Gradually this apparatus was transformed until it consisted of two definitely related parts, the kettle, or body of the still, and the removable head, which, besides closing the kettle, also acted as a condensing device on account of its exposure of a large surface to the air. Further improvements were made from

195

time to time, until the apparatus came to consist of a still body with a detachable head, to admit of the introduction of the material, and of a condensing worm or tube surrounded by flowing cold water. The highly efficient modern still embodies in a more elaborated adaptation the essential principles of this crude apparatus.

Along with the development of the necessary apparatus there have grown up in different parts of the world many large and small industries founded on volatile-oil production. From the small stills formerly used in making essences or spirits for use in the home for medicinal, condimental, or perfumery purposes from herbs gathered wild or grown in the garden, there have come the extensive perfumery industries of southeastern France, the attar-of-rose industry in Bulgaria, the peppermint and turpentine industries in the United States, and the other many and varied phases of the great industry of volatile-oil production.

The present centers of activity in this branch of manufacture have become established where they exist through a favorable combination of conditions, including the adaptation of soil and climatic conditions to the needs of the plants concerned and suitable labor conditions. In southwestern France a general perfumery industry of great importance, based on the production of lavender, cassie, rose, violet, and other perfumery plants, has grown up. The attar of roses from Bulgaria and Turkey, the rose-geranium oils from Algeria, Reunion, and other French colonies, the lavender and other essential oils from England, and the citrus oils from Italy, as well as the lemon-grass, citronella, vetiver, and other volatile-oil and perfume-producing products from India, may be mentioned as important industrial products. In the United States and in Japan the production of peppermint oil and its products constitutes an important industry. In many instances introduced plants are used: in others, native species, usually brought under cultivation, form the basis of production.

The growth of the volatile-oil industry has been most rapid in late years in Germany and France, due in part to the opening up of remunerative lines of work by pioneering scientific workers and in part to the greater demand for these products by the manufacturers of those countries. Although volatile oils find much use in a medicinal way, the greatest demands come from the makers of perfumeries and of flavors. As a result of scientific research along the lines of perfume chemistry, not only has a great field for commercial activity been discovered but scientific knowledge itself has been greatly enlarged. This mutually helpful relation between science and commerce has been conspicuously developed in France and Germany, but to only a relatively slight extent in this country. In view of the increasing importance of this class of products to American commerce, it seems highly advisable that steps be taken to investigate the possibilities of our country in this direction. With our great range of latitude and variety of climate and soil, the conditions naturally favorable to the production of such oils and perfumes should be available. Other questions, such as labor and transportation facilities, must be considered. It is probable that by careful, scientific study of the situation the way may be opened for the development of somewhat extensive industries based on the growing and manufacturing in this country of volatile-oil products now either imported or neglected. These industries are already represented by the peppermint, spearmint, and wormwood products grown in New York, Michigan, Indiana, Wisconsin, and other States of the upper Mississippi Valley.

AROMA OF PLANTS.

NATURE OF ODORS.

Of the countless numbers of plants in the vegetable kingdom, a large percentage possesses peculiar aromatic odors, by means of which the plants may ofttimes be characterized. The substances which impart these peculiar odors to plants consist of mixtures of compounds oily in character and of a volatile nature; hence the designation " volatile oils."

It may be generally stated that all plants which in the growing condition give off a pronounced odor or which produce this odor when the leaves or flowers are rubbed between the fingers contain an essential oil. However, this must not be construed to mean that all volatile oils must necessarily be derived from plants which possess an odor, there being plants which do not possess the oil pre-formed in the tissues, but which through the interaction of constituents in the plant under proper conditions yield a volatile oil. A common example of this class of plants or plant products is the bitter almond, which yields the bitter-almond oil of commerce by maceration of the ground kernels with water, the oil formation taking place during maceration.

The aroma of plants is not necessarily due to volatile oils, there being other odor-bearing substances which, while distinctly aromatic, are not of an oily character. Reference is here made to plants and plant products which, while not possessing any odor during the growing period, develop very fragrant odors after harvesting and drying. An example of this class is the vanilla bean of commerce, which in a green condition is odorless but which when properly cured develops the characteristic fragrant vanilla odor. In this case, according to Lecomte,[a] a glucosidal body in the plant, coniferin, is

---

[a] Lecomte, Henri. Comptes Rendus Hebdomadaires des Séances de l'Académie des Sciences, vol. 133, 1901, p. 745.

hydrolized during the curing process by plant enzymes or ferments to the compound coniferyl alcohol, which in turn is oxidized by oxydase to vanillin. In this case a characteristic odor is perceptible, yet no volatile oil can be separated from the plant. A fuller discussion of this class of substances will follow.

With only a few exceptions it may be stated that volatile oils exist in the tissues of a plant as minute globules, sometimes inclosed in cells but in some instances in enlarged cavities so conspicuous as to be seen without the aid of a lens or a microscope. By a careful examination of the leaf of a peppermint plant, especially at the time of blossoming, tiny glistening particles of oil are clearly discernible. The close scrutiny of the peel of a lemon or an orange discloses to view small, circular oil glands under the epidermis, imparting to it much of the characteristic roughened appearance. Such seeds as cloves, fennel, and anise contain oil passages directly below the epidermis surrounding the endosperm or embryo of the seeds.

The volatile oils in plants do not represent simple substances but are complex mixtures of numerous aromatic compounds which possess a definite chemical composition. However complex the composition of an oil may be, usually one constituent seems to impart the characteristic odor and stands out conspicuously. Generally this constituent attracts attention as the odor bearer of the plant or oil.

The substances which supply the aroma to plants or to essential oils may be resolved by chemical classification into several groups of organic compounds, namely, hydrocarbons, acids, alcohols, esters, aldehydes, ketones, oxids, phenols, and sulphur compounds.

Volatile oils with but few exceptions contain constituents which belong to two or more of the above-mentioned groups of organic compounds. Although each of the groups may contribute to the complex odor of a plant or of a volatile oil, usually compounds exist in the oil which seem to the observer to be especially agreeable and fragrant. The bearers of these pleasant odors which are so apparent even in complex mixtures are for the most part either ester-like or alcoholic in character. It is not unusual, however, that aldehydes, ketones, or phenols play the rôle of odor bearers in a few oils or plants, as, for example, the principal odorous constituent of lemon oil, which is the aldehyde citral, while the pronounced odor of pennyroyal oil is chiefly due to the ketone pulegone. The strongly aromatic odor of thyme is attributed to the phenol called thymol, while sulphur compounds are largely responsible for the aroma of the mustard oils.

Thus it may be perceived that while esters and alcohols impart agreeableness to the majority of oils, there are exceptions, as already stated. Such oils as peppermint, lavender, wormwood, rose, geranium, ylang-ylang, orange flower, and numerous others owe their fragrance to alcohol or ester compounds. or to both. since these compounds are usually found accompanying one another in the oils. Owing to their particularly agreeable fragrance. the esters and the alcohols form a class of the so-called desirable constituents.

Esters represent a group of constituents which are formed by the interaction of alcohols and plant acids (esterification), an ester resulting by the elimination of water in the reaction. Almost invariably these esters possess a pleasant odor and convey the characteristic mellowness and fragrance to many of the essential oils from plants. Indeed, a number of oils are valued according to the percentage of esters which they contain. The largest number of pleasant-smelling esters usually occur in oils as formates, acetates, or butyrates. the acetic-acid esters prevailing. The oil of lavender flowers. for instance. owes its agreeable aroma to the acetic-acid ester of the alcohol linalool or to linalyl acetate. The oil is valued according to the percentage of linalyl acetate which it contains. although the free alcohol linalool also exists in the oil. In this connection it may be mentioned that the ester menthyl acetate imparts fragrance to peppermint oil, menthol being also an important constituent in this case.

Another striking example of an ester compound as the odor bearer of an oil is the methyl ester of anthranilic acid. which carries the odor of orange flowers. Further examples are not necessary to emphasize the importance of esters and alcohols in determining the aromatic value of oils or plants.

In view of the fact that certain constituents may be classed as odor bearers. the desirability of these constituents in volatile oils being evident, attention should be given to the possibility of increasing this class of substances by proper conditions of climate and cultivation.

LOCALIZATION OF ODORS.

Volatile oils, although found in all parts of plants. are localized more or less generally in certain portions. The leaves. possibly on account of their extensive area. often carry a large proportion of oil. In many plants. indeed. the leaves serve as the chief source of the oil. Mention may be made here of the oils obtained from leaves of such plants as the eucalyptus, bay, wintergreen, pine, lemon grass, citronella. and ginger grass. On the other hand. in some plants the oil is obtained principally from other parts. the leaves possessing little or no odor. as in the oil-yielding roses.

The flowering tops of aromatic plants as a rule yield oils of rich aroma. excelling the oils produced from any other portion of the plant. The exquisite bouquet of such oils as rose. lavender, cassie. orange flower, and ylang-ylang is well known, all of these oils being obtained from the flowers or flowering tops.

The fruit oils occupy a position of no little importance, representing an industry by themselves. The principal oils from the citrus fruits are obtained from the lemon, sweet orange, bitter orange (petit grain), and bergamot. In all of the above fruits the essential oil is contained in the peel of the fruit from which it is obtained.

Many of the various seed oils are very important commercially, being employed largely as perfumes and medicinal agents. Among the seed oils derived from the order Umbelliferæ (parsnip family) which possess especial value may be mentioned caraway, anise, fennel, and coriander. Other seeds yielding oils of commercial import are cardamom, American wormseed, mustard, bitter almond, peach, and apricot seeds.

In addition to the above and playing an important rôle in volatile-oil production are the bark and wood oils, the former being represented by such oils as sassafras, canella, and cinnamon. The wood oils comprise such oils as sandalwood, copaiba, and cedar, while from the woods indirectly are obtained several essential oils of value, namely, oils from oleoresins, as turpentine, copaiba, elemi, California turpentine (*Pinus sabiniana*), and Oregon balsam oil.

There are comparatively few root oils, the chief examples being valerian, snakeroot, and sassafras oils.

The aerial portion of the plant serves possibly more extensively for the extraction of volatile oils than any other of the plant parts mentioned. Peppermint, spearmint, and wormwood, from which oils are now produced commercially in this country, are typical instances.

DEVELOPMENT OF AROMA.

The development of the aroma in a plant is conditioned by the interaction of several important factors. It is generally accepted that a close relationship exists between the growth of the plant and climatic factors, such as heat, light, and moisture, and it seems clear also that these conditions play an important part in the formation of the aroma and materially influence its quality. The effect of climate upon the quality of the aroma is clearly shown by the varying fragrance of the oils produced by plants of the same species when they are grown in sections having a wide diversity of climatic conditions. Continuous sunshine, which may be a factor in the development of fragrance in one plant, may possibly exert a reverse action upon another in which the formation of the chief odoriferous constituents is not directly favored by the action of light. Usually, however, sunshine is a favorable agent for the production of delicate aromas, while, on the other hand, cloudiness or darkness has a tendency to lower the production of aromatic substances by the plant.

An abundance of moisture is required for the growth of certain plants and also for the development of aroma. This is especially true of plants whose habitat may be aquatic or subaquatic: in this case dryness becomes a direct hindrance to growth and likewise lessens the activity of the metabolic processes taking place within the organism.

On the other hand, many plants are especial lovers of dryness, particularly such as inhabit the western arid tracts and deserts. These excessively dry regions are not devoid of plant life; neither are they wanting in plants possessing odors. The sages are excellent examples of sturdy growers on dry lands, and many are decidedly aromatic, producing oils of excellent quality.

In both of the above extreme cases, coupled with the dryness or moisture, an abundance of sunshine is usually conducive to the formation of volatile oils in plant organs.

A typical example may be mentioned in the case of lavender. This highly fragrant oil is derived from the plant *Lavandula vera*, which grows for the most part in France and England and is much influenced by such factors as soil, dryness, moisture, altitude, and sunshine. Oils which possess the highest percentage of the odor bearer, linalyl acetate, are usually produced from plants grown on mountain slopes.

Lamothe[a] states that the finest grades of lavender plants of the Dromé region are grown at the highest altitudes (2.500 feet) in the mountain districts. Plants grown on the lowlands of these mountains have been found to be decidedly inferior. Most light soils are well suited to the growth of lavender, but those of a heavy or soggy nature should be avoided.

The lavender produced in the Mitcham district of England is generally considered to have the most agreeable fragrance. In England the conditions are decidedly different from those occurring in France, both with respect to soil and altitude. A chalky soil seems to be best adapted to the growth of lavender in the Mitcham district. The plant is, however, also grown profitably in the vicinity of Bournemouth, Dorsetshire, where the soil consists of sand and clay, with more or less peaty humus.[b] Fungous growths, it is stated, harm lavender where the drainage is not perfect. An abundance of humidity and sunshine is also considered necessary by the English growers.

Although it is generally conceded that the English lavender oil is the most fragrant, this property is attributed by Gildemeister, Hoffmann, and Kremers[c] to the invariably low ester content of the oil,

---

[a] Lamothe, M. L. Bul. Roure-Bertrand Fils., October, 1908, p. 33.

[b] Pharmaceutical Journal, vol. 83, 1909, p. 532.

[c] Gildemeister, Eduard, Hoffmann, Friedrich, and Kremers, Edward. The Volatile Oils, p. 606.

and their findings are further substantiated by Kebler[a] and by Parry.[b]

In the United States the cultivation of lavender has not advanced to any extent. However, in view of the fact that certain regions of the United States possess climate, soil, and other factors practically similar to those of the lavender-producing regions of France and of England, it does not necessarily follow that lavender may not be grown profitably in America.

The nature of the soil through its physical and chemical properties offers an important variable condition likely to affect the metabolism of the plant, and consequently the constituents elaborated by it. Experiments upon peppermint by Charabot and Hebert[c] seem to indicate that soils supplied with commercial fertilizers produce plants yielding oils superior in esters or odor-bearing compounds, the esterification of menthol in the plant seeming to be favored. Peppermint grown by the writer upon a soil rich in organic matter, a black loam, produced an oil noticeably richer in menthyl acetate than peppermint grown upon a clay loam. The existing conditions of climate were possibly also instrumental in bringing about this result.

Seasonal changes have also a marked effect not only upon the quality but also upon the quantity of oil produced by a plant. A plant distilled at its flowering period during one season may produce a certain yield of oil of certain quality, and in the following season, which may be entirely different, it may produce a much higher or lower yield of oil either superior or inferior in quality.

The agents already enumerated are instrumental in bringing about certain chemical changes in the composition of the oil in the cells or tissues of the living plants which contain the oil already formed. There is, however, another group of plants which, though not possessing the oil already formed in the plant tissues, do possess certain basal constituents from which the volatile oil is formed. These constituents usually belong to a class of plant constituents known as glucosids, which break down by hydrolysis into a sugar, generally glucose, and some other compound. The "other compound" which is formed by this hydrolysis in the case of some glucosids is volatile and constitutes the volatile oil from the plant.

Very common examples of plants with glucosidal bodies which yield a volatile oil are wintergreen and sweet birch. The leaves of the wintergreen and bark of the sweet birch contain the glucosid gaultherin, which under proper conditions of hydrolysis yields methyl salicylate and glucose. Methyl salicylate in this instance

---

[a] Kebler, L. F. American Journal of Pharmacy, 1900, p. 223.

[b] Parry, E. L. Chemist and Druggist, 1902, p. 168.

[c] Charabot, A., and Hebert, A. Bulletin du Jardin Colonial, vol. 27, 1902, 3d ser., pp. 224 and 914.

represents the volatile oil of wintergreen. In order to effect this hydrolysis of the glucosid in wintergreen or sweet birch, the material is simply macerated with water. A reaction immediately begins, assisted by the plant ferments, which act as catalysing agents, with the formation of the volatile methyl salicylate and glucose, as follows:

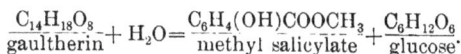

$$\underset{\text{gaultherin}}{C_{14}H_{18}O_8} + H_2O = \underset{\text{methyl salicylate}}{C_6H_4(OH)COOCH_3} + \underset{\text{glucose}}{C_6H_{12}O_6}.$$

If after the reaction is complete the fermented material is put into a distilling apparatus, the volatile oil of wintergreen and sweet birch may be distilled as a colorless oil with the characteristic wintergreen odor so commonly known.

In addition to the two plants mentioned containing glucosidal substances which split up into a volatile oil and a sugar, the ordinary bitter almonds and peach, apricot, and prune kernels may be mentioned. These kernels contain the glucosid amygdalin, which when hydrolyzed yields benzaldehyde, hydrocyanic acid, and glucose, as follows:

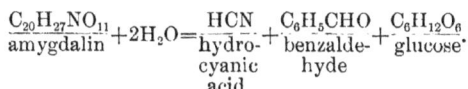

$$\underset{\text{amygdalin}}{C_{20}H_{27}NO_{11}} + 2H_2O = \underset{\substack{\text{hydro-}\\\text{cyanic}\\\text{acid}}}{HCN} + \underset{\substack{\text{benzalde-}\\\text{hyde}}}{C_6H_5CHO} + \underset{\text{glucose}}{C_6H_{12}O_6}.$$

Therefore, when the ground kernels are macerated or hydrolyzed in the presence of water and then distilled, the ordinary volatile oil characteristic of bitter almonds and of peach, prune, and apricot kernels, is obtained.

These kernel oils are in every way identical, just as the oils of wintergreen and sweet birch are practically identical, the former consisting chiefly of hydrocyanic acid and benzaldehyde and the two latter of nearly pure methyl salicylate.

One other example of an oil produced by fermentation is the oil of mustard seeds. These seeds contain the glucosid sinigrin, which likewise suffers hydrolysis when ground seeds are macerated in water, producing the volatile oil of mustard (allyl iso-sulphocyanid), glucose, and potassium acid sulphate, according to the following reaction:

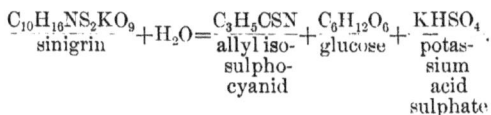

$$\underset{\text{sinigrin}}{C_{10}H_{16}NS_2KO_9} + H_2O = \underset{\substack{\text{allyl iso-}\\\text{sulpho-}\\\text{cyanid}}}{C_3H_5CSN} + \underset{\text{glucose}}{C_6H_{12}O_6} + \underset{\substack{\text{potas-}\\\text{sium}\\\text{acid}\\\text{sulphate}}}{KHSO_4}.$$

The fermented mixture readily yields the volatile oil by distillation with steam. The medicinal action attributed to mustard seeds is due to the mustard oil developed in the reaction mentioned. This oil. however, is not formed until the mustard is brought in contact with water. thus enabling the vegetable ferment to hydrolyse the glucosid, with the results specified.

These instances are cited here simply to make clear the fact that not all volatile oils preexist in plants and that some of our most valuable oils are obtained from plants entirely devoid of odor, which, however, develops when the proper conditions are supplied. The number of these special cases is comparatively few when we consider the vast number of plants which contain volatile oils existing as such in their tissues and depending for their development in the plant only on conditions of growth and nourishment.

EXTRACTION OF AROMA.

For the separation of the aromatic principle from a plant, several methods are in vogue, depending for their efficiency and practicability largely upon the nature of the odors to be extracted. The properties of the various odorous substances are such that in order to separate them in their entirety only such methods can be applied as will bring about the least possible change in the fragrant constituents. Because of the facility with which certain aromatic principles undergo change it is necessary at times to extract the perfume without exposing the materials to high temperatures and to other conditions which would tend to change their chemical nature. For this reason several methods are employed at the present time for the extraction of volatile oils and perfumes, each of which possesses advantages and disadvantages.

The following general methods find application in commerce for the separation of the odoriferous principles from plants and plant products: (1) Solution, (2) expression, and (3) distillation.

SEPARATION OF PERFUMES BY SOLUTION.

The method of solution as applied in practice is subdivided into three modifications, viz. by volatile solvents. by liquid fats. and by solid fats.

EXTRACTION WITH VOLATILE SOLVENTS.

The method of extraction with volatile solvents, such as ether, chloroform. benzene. petroleum ether. acetone. etc.. is adaptable only to flowers. because of the comparatively small quantity of other kinds of extractive matter soluble in any one of these solvents. The method would be very impractical for the extraction of perfumes or oils from a whole plant or from the leaves of a plant. since whole plants or plant parts other than flowers contain considerable other matter besides the essential oil soluble in these solvents.

195

The method employed commercially for the extraction of odors by means of these volatile solvents embodies a process known as continuous extraction. By this method the solvent, after percolating through flowers and carrying with it in solution the odorous constituents, is heated in a proper receiving vessel and the vapors condensed and utilized further for extracting any remaining odor. The advantage of this method is the small amount of solvent necessary for extraction and the continual percolation of fresh solvent through the material.

The accompanying illustration (fig. 1) represents an apparatus used for this purpose, which consists chiefly of the percolator, the receiving vessel, and the condenser.

The percolator, B, in the bottom of which is placed a circular screen, is charged with the flowers to be extracted, and the removable cover, F, is attached by means of clamps, as indicated. A heavy gasket of cotton wicking or asbestos (previously moistened) or rubber is placed between the cover and the percolator to insure a tight connection. To the bottom of the percolator at H is attached the receiving vessel, A, and the hot water steam bath, D, by means of a screw union. Into the cover, F, is fitted a perforated rubber cork, through which passes a glass tube, K. The glass tube, K, is further connected with the condenser, C, by means of a perforated rubber stopper. The condenser may be of the single-tube or worm variety, the former being preferable. The tube K is of glass for the purpose of enabling the operator to observe the rapidity with

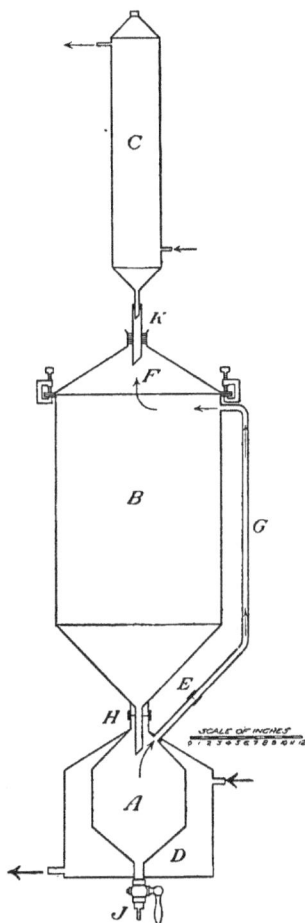

Fig. 1.—Continuous extraction apparatus. A, Receiving vessel; B, percolator; C, condenser; D, bath; E, union; F, cover; G, tube; H, union; J, drain cock; K, glass tube.

which the condensation of vapors is taking place. After pouring the solvent through the condenser and into the percolator, heat (preferably steam or hot water) is applied to the bath, D. The steam is passed through the bath, D, in the direction indicated by the arrows. The solvent which has percolated through the flowers in B is vaporized and driven up through the tube G (which should be covered with asbestos to prevent radiation) and into the percolator, thence into the condenser, where the vapors are condensed and drop back into the material. A continuous extraction is thus obtained with a minimum quantity of solvent.

For the final recovery of the solvent from A, the apparatus, after cooling, is disconnected at H and a screw cap attached to the neck of A. The tube G is disconnected at the union E, which may be connected with the condenser in proper position, and heat applied to D. The excess of the solvent is completely recovered in this manner, the resultant oil or perfume being drained off by opening the cock, J.

The chief disadvantage of an apparatus of this type is its narrow field of usefulness, which is practically restricted to the separation of perfume from flowers. When this apparatus is used for the extraction of other parts of the plant which may contain aromatic substances, the oil is liable to be contaminated by resins, waxes, etc., which would be extracted with the perfume by the solvent used. In order to purify further the crude oil obtained, steam distillation must be resorted to, in which case the delicate quality of the perfume obtained by the cold extraction would probably suffer slight changes induced by the steam.

EXTRACTION WITH LIQUID FATS.

The process of extraction with liquid fats is comparatively simple and depends upon the ability of a liquid, fatty oil to absorb the odors from flowers. For this purpose olive oil, lard, or other bland fixed oils may be advantageously used. The oil is placed in a kettle or vat (preferably porcelain lined) and heated to a temperature of 40° to 60° C.; the flowers to be extracted are then introduced either directly into the fatty oil or inclosed in coarse bags and suspended in the fat. The material is maintained at this temperature for a time varying from one-fourth to one and one-half days, when the mixture is either drained to remove the flowers or the bags are removed and expressed and recharged with fresh material. In this manner a perfumed oil is produced from which the perfume may be extracted by shaking out with strong alcohol, in which the odor is soluble and the fat insoluble. The fatty oil, which still retains traces of the flowery fragrance, may be used for further extraction of the same flowers.

This method of maceration in liquid, fatty oil is carried on to some extent in the perfume gardens of southern France and Germany, where perfumed oils are largely manufactured from such flowers as rose, jasmine, violet, tuberose, cassie, etc.

The extraction by maceration is advantageous because of its ease of operation and manipulation, but owing to the fact that heat is necessary for the rapid absorption of the perfume, another method in which the fat is used as a cold absorbing medium has been devised and used.

EXTRACTION WITH SOLID FATS.

The process of absorption of perfumes in cold by means of fats, the "enfleurage" process, has long been used for the extraction of the more delicate odors, and is possibly more universally used than any other process for the preparation of certain flower odors.

The great avidity with which some solid fats absorb aromatic substances is the basis of the method. Odors of nearly every description are absorbed by neutral solid fats when the latter are placed adjacent to or in contact with the odoriferous substances.

The enfleurage process, which is based upon this peculiar property of fats, was originally carried out by spreading freshly picked flowers upon a thin layer of lard spread upon glass plates, the flowers being allowed to remain in contact with the lard until exhausted, when the apparatus was charged with fresh flowers. In this manner a perfumed pomade was produced containing the natural odor of the flowers.

For effecting a separation of the perfume from the solid fat, which is desirable in some cases, advantage is taken of the comparative insolubility of the fat in strong alcohol and the ready solubility of the perfume. Therefore, in preparing the pure perfume, the perfumed pomade is thoroughly and repeatedly agitated with alcohol, an alcoholic extract or perfumed essence resulting. This resulting extract is sometimes employed as such for producing delicate scents. In order to obtain the pure oil from the alcoholic extract, the alcohol is evaporated carefully in a vacuum, the concentrated oil or perfume of the flowers remaining. These concentrated oils, although often rather unpleasant in odor in extreme concentrations, produce an exquisite aroma when diluted.

The crude process of enfleurage just mentioned has been largely modified in recent years in order to promote rapidity of operation, to protect against loss of odor by nonabsorption, and to obviate the actual contact of the flowers with the lard. When the flowers are in actual contact with the lard there is a tendency toward the absorption of undesirable substances.

A practical apparatus of this nature (fig. 2) consists of a box, *H*, about 2 feet square and 6 feet high, so constructed as to be practically air-tight. In the lower portion of the box, which is supported about 2 feet above the floor, is placed a layer of sponges, *G*, or other porous material capable of holding moisture. The bottom of the sponge tray may be constructed of light copper gauze or brass gauze to permit the free access of air. Directly above are located the flower trays, *A*, *B*, *C*, *D*, and *E*, which also have brass or tinned-iron screens of rather coarse mesh for bottoms. The sides, fronts, and backs of trays may be of wood. The trays may readily be placed in or taken out of the absorption box when refilling is necessary. Immediately above the flower trays are located a series of glass plates so constructed that they may be readily taken from the box and replaced.

The absorbing medium, lard or other solid fat, is spread in a layer about one-half inch in thickness upon each glass plate, which is placed in its proper position. The front portion of the apparatus must be supplied with a tight-fitting door (not shown in the illustration) capable of being opened or removed to admit of charging and discharging the fat and flowers. When the flower trays have been charged with the freshly picked flowers and the door closed firmly a current of air is made to pass upward through the sponges and the flowers and the lard-laden tray, a more efficient circulation being produced by the alternating arrangement of glass plates. The odor-bearing air as it passes over the lard readily surrenders its perfume, which can be subsequently extracted from the lard. A small fan may be placed at the top of the apparatus or a blower at the bottom to produce the required movement of the perfume-laden air. The current should be regulated so that absorption is completely effected in its upward journey.

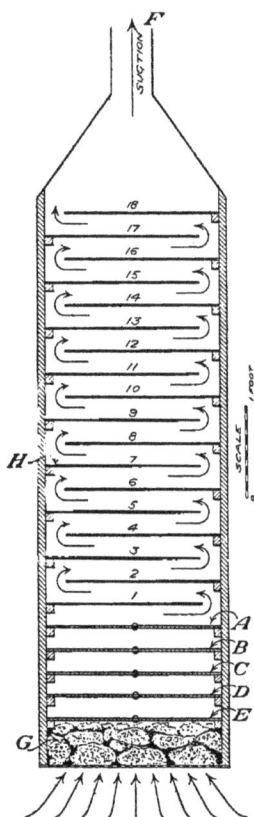

When retained in fresh condition, flowers hold their aroma and even secrete perfume for a longer period of time than if allowed to wilt and dry; hence the moistened sponges in the bottom of the apparatus. Some flowers are even known to continue to secrete perfume if left in moistened air. The air drawn through the apparatus is moisture laden and therefore produces the best yield of perfume from the flowers.

FIG. 2.—Apparatus for treating flowers by the enfleurage process. *A*, *B*, *C*, *D*, *E*, Flower trays; *F*, exit; *G*, sponge tray; *H* (1–18), glass plates.

The operation of the above contrivance may be continued with only such interruption as is required for recharging with fresh flowers when practically all odor has been drawn off. After the lard has been thoroughly charged, the perfume held in solution is

separated by a thorough agitation of the pomade with strong alcohol, preferably by means of a shaking or churning device in which
the pomade is continually agitated and beaten in order to expose the
largest surface possible to the solvent action of the alcohol. There
results from this extraction operation an alcoholic extract of the
flowers which possesses the natural odor to a very high degree. Because of the fact that no heat is necessary, the resulting extract is far
superior to an extract prepared by the process of heating with liquid
fats.

It is to be remembered, however, that the yield of perfume from
some of the more delicate flowers, such as violet, cassie, tuberose,
jasmine, etc., is rather small, which accounts largely for the exceedingly high prices of the extracts or pomades of these flowers.

Usually it is impossible to extract the odor from the pomade completely, even when extracted successively with fresh portions of
alcohol. The fat after extraction still retains the characteristic
aroma and may be used in this form or may be again spread upon
the glass and utilized for further absorption from the same kind
of flowers.

The amount of labor required for this work is necessarily large
when the fact is taken into consideration that the flowers require hand
picking. The time consumed by the entire process from the picking
of the flowers to the finished extract is also very considerable. However, the quality and, consequently, the prices of these exquisite odors
usually offset unfavorable conditions of labor and time in regions
where this industry is carried on commercially.

SEPARATION OF PERFUMES BY EXPRESSION.

Another class of volatile-plant products already cited is so localized
in the plant as to admit of the extraction of the oil by a different yet
extremely simple process. The class of products referred to includes
the citrus fruits, namely, the lemon, orange, bergamot, and other
related fruits. Owing to the fact that the oil contained in these
fruits is deposited in the outer portion of the peel and is therefore
very accessible, the method of expression is peculiarly adapted to the
citrus fruits and products.

There are several methods applicable to the extraction of the oil
from the peel of the lemon, orange, and bergamot, all of which, however, embody the same principle, namely, the rupturing or breaking
of the glands containing the oil and the collecting of the oil after it
has been released.

In the method known as "écuelle à piquer," the rinds of the
lemons are rubbed in hollow cups (écuelle, fig. 3) lined with sharp
points, which lacerate the oil glands and allow the oil to exude.

This method has been largely displaced by the simple expression of the oil.

Owing to the ease with which the peels of the fruits liberate the oil, a method of expression is applied very conveniently to the separation of the oil. Usually the peels from half sections of the fruit are turned inside out and pressure brought to bear on the outer surface in such a manner as to rupture a large majority of the oil vessels. The oil thus liberated is collected upon a sponge, which absorbs it and from which it is subsequently squeezed. By this method, known as the "sponge method," the larger part of the oils of the lemon, the orange, and the bergamot is extracted, the operation being carried on usually at night, when other activities in the fruit work are at a standstill.

Expression by the sponge method is far from complete because of inability to bring pressure upon every portion of the peel; hence, after the "hand-pressed" oils, which are generally conceded to be the best grade, are obtained the peels are placed in a power press or in a crude still and the remaining oil is separated. This latter forms a secondary oil of commerce, generally considered to be much inferior to the sponged oil.

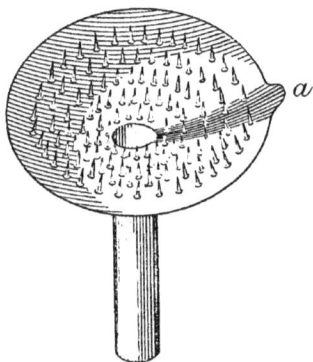

FIG. 3.—Écuelle for lacerating the oil vessels in the peels of oranges, lemons, etc. a, Draining lip.

The use of a mechanical device for rupturing lemons and bergamots and for expressing the oil from them has been introduced into some producing districts of Europe. However, only a small percentage of the oils is extracted in this way, the sponge system being most usually adopted.

Whether the process of steam distillation, which will be discussed later, if somewhat modified would produce a grade of oil equal to the hand-pressed oil is doubtful. At any rate, the oil containing only traces of compounds capable of decomposition at the usual temperature of steam, it should not be greatly inferior, production by this method would be easier, and its cost would be materially less.

SEPARATION OF PERFUMES BY STEAM DISTILLATION.

A simple still, which consists essentially of three parts, the still body, the condenser, and the receiver, with a suitable means of applying direct heat to the still body, containing material suspended in water, was used early in the eighteenth century. Even at the present time many smaller distillations are still carried on with this form of apparatus. The chief disadvantage of this type of still lies in the fact that the heat, being applied directly, has a tendency to char or burn the materials adjacent to the bottom, and thus appreciably affect the quality of the aromatic product distilled over.

This method has been largely superseded in modern times by distillation with steam, the principles of which depend upon the property of the steam as it passes through the charged apparatus to carry with it the volatile portion of the plant in the form of vapors, which are condensed, together with the excess of watery vapor, and deposited in the receiving vessel. The three steps in the process are (1) the distilling, (2) the condensing of the vapors, and (3) the collecting of the oil. Even though the boiling points of the volatile oils separated by distillation from plants may be considerably higher than the temperature of steam, the odors are readily liberated by the passing steam and carried over.

APPARATUS.

The apparatus required for the three processes which collectively constitute steam distillation is of comparatively simple construction, consisting of (1) a still, (2) a still head (cover for body), (3) a condenser, and (4) a receiver.

The body of the still, or the receptacle in which is placed the material from which the oil is to be extracted, gives best results when cylindrical in form and may be constructed of various materials, preferably copper. However, some stills are made with wooden bodies. Galvanized iron heavily tinned on the interior is a suitable material, principally because of its cheapness and durability. The still may be constructed of any size desirable, provided the other parts, the condenser and the receiver, are in proportion, depending upon the amount of material to be used and the extent of production desired.

In figure 4, *A* represents the still, *B* the still head, or cover, *C* the condenser, and *D* the receiver. Through the side of the still at the point *E* passes a galvanized steam pipe from three-fourths to 1 inch in diameter, extending downward and finally terminating in the middle of the still, as shown by the dotted line. A spigot, *F*, is attached to the bottom of the still for draining the collected water from the apparatus. About 3 inches from the bottom of the still is placed a coarse screen, *H*, fastened to a wooden frame, which acts as a support for the herb or plant part to be distilled. Encircling the top of the still is an iron collar, which may be conveniently constructed of angle iron, to which the copper or the metal is securely attached.

The still head, or cover, *B*, is of the same material as the still and is slightly conical in shape, with an exit tube terminating in a union, at which point connection may be made with the condenser. Around the periphery of the cover is securely fastened a flat collar of iron of the same diameter as the angle iron used on the top of the still, so that with the cover in place the two will exactly coincide.

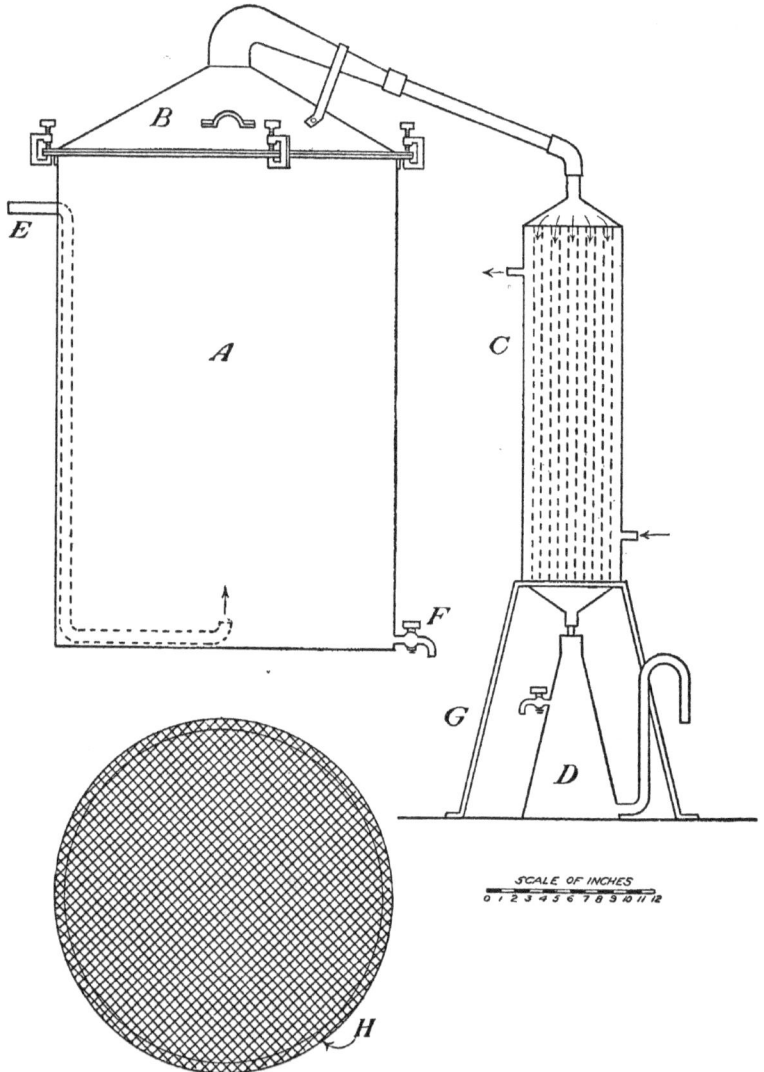

FIG. 4.—Distilling and condensing apparatus. *A*, Still; *B*, still head or cover; *C*, condenser; *D*, receiver; *E*, steam pipe; *F*, spigot; *G*, tripod; *H*, screen.

The condenser, *C*, as shown in figure 4, consists of a group of tubes (inside diameter one-half to 1 inch, depending upon the size of the condenser) surrounded by an outside jacket fitted with an inlet tube at the bottom and an outlet tube at the top, to enable cold water to pass continually through the condenser in an upward direction. The condenser is attached to the still by means of the union joint, as illustrated.

The tripod, G, acts as a support for the condenser while the apparatus is in operation and also while the still is being charged or discharged. Under the bottom opening of the condenser is placed a receiver, D, of copper, with a goose-neck siphon tube extending from the bottom to within 2 inches of the top. On the side opposite the siphon tube is fastened a small brass spigot to admit the removal of the oil from time to time.

For the generation of steam, if a source is not otherwise available, a small boiler, such as is illustrated in figure 5, may be conveniently used. A small boiler, A, of light boiler iron fitted with about a dozen flues is capped by the cover, B. Other usual accessories are attached, viz, water gauge, C; pop valve, D; water gauge, E; and steam outlet, F. The boiler may be preferably set upon a gasoline stove or an open-fire stove or on a tripod with an open fire beneath. The pop valve may be set at about 8 to 10 pounds, no greater pressure being necessary. To replenish the water in the boiler a funnel tube attached to the pop valve may be used. Connection to the still is made most conveniently by the attachment of a short piece of rubber steam hose

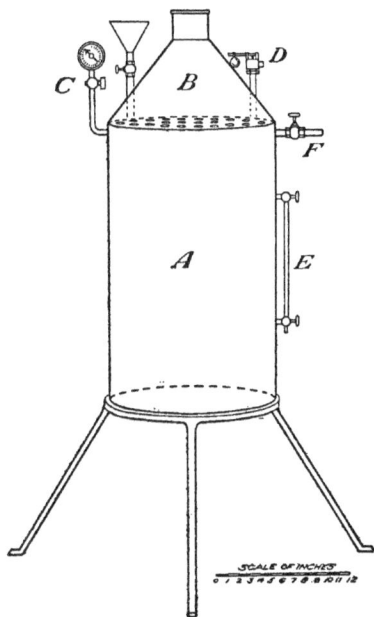

FIG. 5.—Steam generator. A, Boiler; B, cover; C, steam gauge; D, pop valve; E, water gauge; F, steam outlet.

to F, as this admits a ready detachment from the still when distillation is completed. A pressure of 5 to 10 pounds of steam is sufficient for ordinary distillation. The size of the boiler may be slightly increased if distillation is to be conducted on a larger scale.

The boiler just described possesses efficiency enough to distill charges of 75 to 150 pounds of herb.

For distillation on a commercial scale a large, stationary, upright boiler may be installed for the generation of steam, or, if convenient, steam may be taken from any high-pressure boiler which may be in use for other purposes. The volume necessary being very slight, indeed, is scarcely perceptible upon the steam gauge.

59647°—Bul. 195—10——4

To charge the still, place the false bottom in the still and pack the herb firmly until completely filled. Place a gasket of asbestos rope, heavy cotton wicking, or other suitable material (previously moistened) around the top of the still. Place the cover upon the moistened gasket and clamp securely with heavy steel clamps. Connect the exit pipe from the top of the still to the condenser by means of the union, as indicated in the diagram. Now conduct steam into the still through the inlet pipe, E, slowly at first, and regulate afterwards so that the distillate passing from the end of the condenser is cold or but very slightly warm. The receiving vessel, D, should be previously filled three-fourths full of water and placed under the exit from the condenser. Likewise, the cold water is started flowing through the condenser, as indicated by the arrow. Frequently the oil may be led from the receiver by opening the cock on the side. However, owing to the siphon tube attached to the receiver, overflowing is impossible, since this tube carries off the water which separates in the bottom of the receiver. To ascertain when distillation is completed a few drops of the distillate as it comes from the condenser are collected in a glass test tube. The appearance of oily globules on the surface readily indicates whether appreciable quantities of oil are still passing over. Usually a distillation is completed in from one and one-half to two and one-half hours.

The advantage of steam distillation over other methods of volatile-oil extraction lies principally in its wide applicability and speed of operation. Most plants or plant parts, with the exception of the flowers in some few cases, may be extracted most readily and most expeditiously and with a minimum amount of labor by the steam-distillation method. The simplicity of the operation is obvious. The removal of the oil is much more complete than by any other process. Furthermore, there is produced as a by-product during the distillation an aqueous distillate which is completely saturated with the oil. The aqueous distillate may in many instances be utilized and sold as an "aromatic water" of commerce, especially in such cases as lavender, orange flowers, rose, etc. The aromatic waters possess excellent odors, largely because of the extreme dilution of the odorous compounds held in solution, and are useful in the perfumery and toilet-preparation industries. When the aqueous distillate from the plant has no marketable value, it may be profitably collected and returned to the boiler. In case of a further distillation of the same plant it will materially add to the yield of oil, since the distillate is a saturated solution of the oil. Many oils are extremely soluble in water. Distillates from oils of this class usually augment considerably the yield of oil when returned to the boiler and transformed into steam and oil vapors.

195

The spent herb, which on a large scale amounts to no inconsiderable quantity, may be used as fuel and the ash used as fertilizer, or it may be scattered upon a field and plowed under as a mulch. In some cases the spent herb serves as a useful stock food, an example of which is the peppermint grown in Michigan.

The advantages far outnumber the disadvantages of the distillation method, the only disadvantage being the possibility of slight decomposition of the ester bodies in some of the more delicate perfumed plants. However, this is only slight and almost negligible in most herbs.

HANDLING OF VOLATILE OILS.

PURIFICATION.

The volatile oil as it comes from the still is in a crude state, being contaminated by volatile substances which are formed during the distilling process by the action of the steam upon the less stable plant constituents, decomposing them into volatile organic substances, which, although trifling in quantity, nevertheless tend to affect the color, odor, and taste of the oil.

The chemical changes taking place in the still are numerous, the more important being oxidation and reduction of some of the constituents of the oil, as well as of the other plant constituents, saponification of the more unstable esters, and resinification brought about by a polymerization of certain plant constituents, all of which aid in forming volatile substances which mingle with the oil.

Although a process of purification is not always applied to these crude oils, it is important and sometimes highly profitable to subject the crude product to a process of rectification. By rectification is meant a redistillation of the oil with steam, this procedure affecting a moderate separation of the undesirable substances which may have been formed. The substances which detract from the odor of the oil are usually left behind in the apparatus as a heavy, malodorous liquid slightly resinous in character. Rectification usually results in a fine, finished product, free from foreign odors, and leaves an oil much more presentable in color as well as in odor and taste.

This process may be conducted in a miniature still built on the same general plan as the large commercial still. The loss in the amount of oil is more than compensated for by the better quality and the increased salability of the rectified oil.

SEPARATION, FILTRATION, AND DRYING.

To separate the oil from the aqueous distillate in the receiving vessel, the portion which has not been separated by means of the stop cock on the side of the receiver is poured into a separating funnel of glass and the heavier liquid drawn off. The oils resulting from different distillations of the same plant are then united and subjected to filtration, which process tends to separate any solid particles or emulsion of oil and water. Filtration is conveniently effected by pouring the oil into a glass funnel which has been fitted with a filtering medium, such as filter paper (an unsized, porous paper) or cotton. When cotton is used as a filtering medium a small tuft may be fitted loosely into the neck of the funnel and oil poured upon it. Usually filtration takes place more rapidly through cotton than through paper and with much less loss. Rapidity of filtration is essential to minimize the possibilities of changes taking place in the oil by oxidation, since the oil is more or less exposed to the action of the air and light while undergoing this clarifying process. Hence cotton is to be recommended.

Just as the water that comprises the aqueous distillate is a saturated solution of the oil, so the oil which floats above the distillate is saturated with water. Usually it is of prime necessity that the moisture be removed from all oils, first, because of the subsequent changes that are likely to occur if moisture is present, and, second, because of the turbidity which water imparts to the oil. Hence, after filtration through cotton the oil should be dried by shaking in a bottle with a dehydrating substance, such as anhydrous calcium chlorid or anhydrous sodium sulphate, preferably the latter, owing to its lack of action upon the constituents of the oils. The crude sodium sulphate (Glauber's salts) may be dehydrated by heating it in a vessel over direct heat, with constant stirring until a dry, grayish powder results. But a small quantity is necessary to abstract the moisture from an oil. After the oil has been dried it is again filtered through a light plug of cotton. A clear and transparent oil finally results, bearing in every way the appearance of a marketable oil.

PRESERVATION.

Many constituents of volatile oils are of such a nature that unless the strictest precautions are observed in storing the oils chemical decomposition takes place, causing them to change in both odor and color, thereby reducing the quality and value. The esters of an oil (combinations of organic acids with alcohols) are very prone to decomposition, as are also many aldehydes and hydrocarbons, which either through saponification, hydration, oxidation, reduction, or polymerization become totally different substances. These chemical processes are usually stimulated by the action of light and air upon the oils. Therefore, in order to guard against these changes and to minimize them as much as possible, the strictest attention should be paid to the proper bottling and storage of the oils.

It is of the utmost importance that all oils should be placed in bottles which are well filled. The absence of air is of the greatest importance in insuring the preservation of an oil. The oxygen of the air, assisted by light, becomes extremely energetic in bringing about some of the changes previously mentioned. It is therefore of import that the oils be kept not only in well-filled, tightly stoppered bottles, but in a dark place. It is sometimes convenient and advisable to use amber-colored bottles in order to prevent the entrance of the actinic rays of light which are so active in causing polymerization. A cool place is also to be preferred for the storage of volatile oils.

All undue exposure of oils to the action of light and air should be avoided as much as possible. It is necessary that an oil from the time it leaves the receiving vessel after distillation or rectification until it is filtered, dried, and bottled should be handled with care and dispatch to insure a product of the best quality and appearance.

GROWTH AND HARVESTING OF PERFUME PLANTS.

CLIMATE AND SOIL.

Up to the present time the cultivation of perfume-yielding plants has not been carried on, even experimentally, over a very large part of the United States, and such work of this sort as has been done is confined to but a few kinds of plants. Until our knowledge along these lines has been very much increased by practical attempts to cultivate this class of products, only statements of probabilities can be made. However, in some cases plant introductions along other lines from the oil-yielding countries of the Old World, together with information as to conditions of climate and soil in those regions, give a basis for surmise in connection with these crops. The wide diversity in climate and soil in different parts of the United States, with the varying conditions of heat, light, and moisture, renders it probable that some portions of the country will be found to be well fitted for the cultivation of the perfumery plants characteristic of the temperate zones. It appears probable that the conditions prevailing in those parts of Europe associated with the perfumery industry can be fairly well duplicated. It will doubtless require much experimental work to find the particular localities best suited to special plants.

It must be borne in mind, however, that not only must conditions of soil and climate be right but that the labor conditions which go with the problem must be met in a practical way. The distance of the point of production and the transportation factors are also important and might be decisive.

Some work on perfumery-plant growing has been carried on in Florida, notably by Mr. E. Moulié, of Jacksonville, whose experience has been distinctly encouraging. Experiments by Mr. S. C. Hood with a number of oil-yielding grasses grown in the testing garden carried on by the Bureau of Plant Industry at Orange City, Fla., give good ground for hope that a number of kinds of plants able to endure a little freezing weather may be cultivated with good results. California and the arid Southwest offer promising conditions for plants which thrive in dry, sunny locations. Michigan, Indiana, and New York are already well known as important centers for the production of peppermint, spearmint, and erigeron oils, while Michigan, Wisconsin, Nebraska, and other States in the north-central part of the country form a most important source of wormwood oil. Doubtless other oil-bearing plants now on trial may be found to do well in parts of the same general section. American wormseed (*Chenopodium* spp.) is distilled in Maryland and southward, and sassafras is distilled in various places, especially in the mountains, from Pennsylvania southward. The oils of wintergreen, sweet birch, spruce, and white cedar are derived from the more northern ranges of the Atlantic slope. The mountainous regions of Tennessee and Kentucky supply wintergreen, sweet birch, and sassafras oils.

It is thus apparent that a number of native and introduced plants rich in volatile oils have obtained foothold on a commercial basis in this country, and there is good ground to hope that products of this general class now obtained from abroad may in time become naturalized here.

GROWTH AND CULTIVATION.

Several methods of procedure with regard to the propagation and cultivation of volatile-oil and perfume-yielding plants are to be followed, depending largely upon the nature and habitat of particular species of plants. Annual plants such as are grown from seeds and which blossom and mature the same year are rather common among volatile-oil plants.

The details of cultivation and handling vary somewhat with the crop grown and are a matter for careful field study. In general, the annuals are either fall or spring sown, depending upon soil and climate, some seeds germinating best if left in the ground over winter, as is the case with pennyroyal. Row culture is advisable in order to secure better cultivation and a consequent freedom from weeds.

Perennials are in some cases grown well from seed, as caraway and wormwood, but in some cases, such as spearmint, peppermint, sage, rose, and lavender, propagation from cuttings or roots is preferable.

The method of handling must be adapted to the particular plant to be grown.

A thorough cultivation of the field is necessary to eliminate all weeds, both between the rows and in the rows themselves. This is of the utmost importance, since weeds, although as a rule not containing any volatile oil, do possess volatile substances which are set free by the steam should the weeds become mixed with the aromatic plant. A contamination of the oil and a depreciation in the aromatic qualities will result unless the material is kept free from weeds and other rank growths.

HARVEST.

Possibly no stage in the cultivation and production of volatile oils from plants is of greater importance than that of the proper harvesting of the crop. It is usually conceded that most perfume plants reach their maximum development as regards odor, both in quality and quantity, at the flowering period. On the other hand, many authorities are of the opinion that as soon as a plant reaches its full flowering period there sets in a gradual consumption of the odorous principles; hence, the harvest should be made prior to this consuming process.

Experiments recently conducted for the purpose of determining the amounts of odorous constituents of several plants present at various stages of development seem to indicate that both the quality and the quantity of the oils vary appreciably during their successive stages of development, but no evidence was obtained to show that consumption of odor took place during flowering. However, it was proved that the odor was developed during the advance in growth and the approach of the flowering period.

Three typical plants were used as a basis of experiment, viz, peppermint (*Mentha piperita*), bergamot mint (*Mentha citrata*), and wormwood (*Artemisia absinthium*), the oil of each of which owes its characteristic fragrance to esters which admit of being measured quantitatively with some accuracy. The plants were grown under like conditions and distillations conducted at three well-defined stages of advancement, namely, (1) before flowering (or while in the budding state), (2) at flowering, and (3) after flowering (or during the fruiting stage).

The effect of successive stages of growth upon the esters and the alcohol only will be considered here, although other constituents, and especially the terpenic compounds, also suffer changes.

195

To picture more clearly the results of the experiments and the changes observed in the oils. tabulations were made as follows:

TABLE I.—*Yield of oil and changes observed in plants at different stages of growth.*

PEPPERMINT (MENTHA PIPERITA).

Stage of growth.	Yield of oil.	Ester content as menthyl acetate.	Alcohol content as free menthol.
	Per cent.	*Per cent.*	*Per cent.*
Before flowering (July 22)	0.23	9.5	31.0
At flowering time (August 21)	.20	14.5	23.6
After flowering (September 25)	.10	24.0	34.0

BERGAMOT MINT (MENTHA CITRATA).

Stage of growth.	Yield of oil.	Ester content as linalyl acetate.	Alcohol content as linalool, free.
	Per cent.	*Per cent.*	*Per cent.*
Before flowering (July 20)	0.32	47.6	7.3
At flowering time (September 22)	.37	55.0	7.3
After flowering (October 14)	.22	52.0	5.5

WORMWOOD (ARTEMISIA ABSINTHIUM).

Stage of growth.	Yield of oil.	Ester content as thujyl acetate.	Alcohol content as thujyl alcohol, free.
	Per cent.	*Per cent.*	*Per cent.*
Before flowering (July 2)	0.19	26.0	14.7
At flowering time (July 14)	.18	32.5	11.7
After flowering (August 4)	.10	47.5	12.0

It is obvious from these results that in two cases, with peppermint and with wormwood. the aromatic quality of the oil, if measured by the percentage of esters. is increased gradually during each stage of growth, the percentage of free alcohol remaining fairly constant. In the peppermint the oil from the "after-flowering" stage was noticeably more fragrant than the oils from the two earlier stages. The yield of oil remains fairly constant up to the last stage, when there is a marked diminution. The plant in the first two stages is very much the same as regards moisture content, while the low percentage of oil from the plant after flowering, when it possesses much less succulency, may be attributed to the consumption of other constituents than the esters and alcohol. This applies to all of the plants which seem to follow the same general course in this respect.

The oils from the bergamot mint disclose a very slight decrease in ester content and alcohol content in the "after-flowering" stage.

The decrease is so slight, however, as not to warrant the statement that a consumption of odor has occurred.

It must be understood that these results are proposed only tentatively and that further experiments will be carried on to prove or disprove the conclusions drawn.

Employing the aforementioned plants as typical examples, the harvest period, in order to attain a maximum yield of oil with a correspondingly high percentage of odorous constituents, should begin as soon as the plant is fully blossomed. A delay of the harvesting until the "after-flowering" stage is reached apparently increases somewhat the quality of the odor, but this increase is largely overbalanced by the decrease in the yield of oil, which is of paramount importance to the grower.

The proper preparation of the material prior to distillation is not to be overlooked, since the quality and the quantity of the oil are varied considerably by improper handling and by partial or complete drying of the fresh plant before it enters the still.

To illustrate this point more clearly, practical instances will be mentioned to show the effect of drying upon the quality and the quantity of the oil from plants. The three examples previously mentioned will be used as a basis for the comparison of the oils from fresh and dry material. In order to obtain a rational and logical means for comparing the oils, fresh, green plants of peppermint, bergamot mint, and wormwood were cut during the height of their blossoming stage. The herb in each case was divided into two equal parts, one half of which was set away to dry and the other half distilled immediately. The oils obtained were later analyzed for the esters and the alcohols, and the results obtained are presented in Table II.

TABLE II.—*Yield of oil and percentages of esters and of alcohols obtained from fresh and from dry plants.*

PEPPERMINT (MENTHA PIPERITA).

Condition of plant.	Date of distillation.	Yield of oil.	Menthyl acetate.	Menthol.
		Per cent.	*Per cent.*	*Per cent.*
Fresh	August	a 1.50	10.5	48
Dry	December	.55	18.0	47

BERGAMOT MINT (MENTHA CITRATA).

Condition of plant.	Date of distillation.	Yield of oil.	Linalyl acetate.	Linalool.
		Per cent.	*Per cent.*	*Per cent.*
Fresh	September	a 1.30	33.0	45
Dry	December	.75	51.8	43

a Calculated from dry weight.

TABLE II.—*Yield of oil and percentages of esters and of alcohols obtained from fresh and from dry plants*—Continued.

WORMWOOD (ARTEMISIA ABSINTHIUM).

Condition of plant.	Date of distillation.	Yield of oil.	Thujyl acetate.	Thujyl alcohol.
		Per cent.	*Per cent.*	*Per cent.*
Fresh	August	0.60	32	41
Dry	December	.44	85	15

These data with respect to the oils from fresh and dry herbs readily illustrate that during the drying of the plants certain factors, assisted by exposure to air and light, undoubtedly bring about chemical changes in the aromatic constituents, which evidence themselves in the final analyses of the oils.

It will be noted that the yield of oil decreases 63⅓ per cent in the case of peppermint, while the percentage of decrease of oil from bergamot mint is nearly 43, and from wormwood about 27 per cent. These marked decreases are in part due to the long period of drying, but they at least show that there is a downward tendency, which is very natural considering the volatility of the constituents.

In all three cases there seems to be an increase in the percentage of esters, with a decrease in the percentage of alcohol, in the dried herb, the chemical changes no doubt being such as to facilitate the production of esters and to break down the alcohols. Apparently the alcohols seem to be more unstable, condensing with the organic acids in the plant under favorable conditions of heat, light, and moisture to form esters. This latter change is especially noticeable in all the oils, the dry-herb oils being considerably richer in esters than the fresh-herb oils, and correspondingly poorer in alcohols.

In order to produce the largest yield of oil from a given quantity of herb, distillation should be made immediately after harvesting. There is no noteworthy advantage in drying or even partially drying the plant, since the longer the time between the cutting and the distilling the more volatile oil will be lost by gradual evaporation or volatilization. Although the quantity of oil capable of being carried off into the air by simple drying seems only trifling, nevertheless, on a large scale the loss would be considerable. The increased proportion of the odoriferous esters in the oils from dry herbs is insufficient to warrant the drying of the plants before distillation, because of the loss of oil encountered during the drying process.

VOLATILE OIL PLANTS OF THE UNITED STATES.

At the present time the number of plants in the United States yielding volatile oils in a commercial way is very small, but the number capable of yielding oils of probable value is correspondingly great. There is, in fact, a large number of odoriferous plants still uninvestigated which should demand consideration. As yet but little research has been undertaken which would tend to increase the number of valuable aromatic plants now being utilized. A study of this particular phase of the subject, coupled with the introduction of foreign species into the United States, should eventually develop somewhat the resources of the country along this important line.

CULTIVATED PLANTS.

The relatively small number of volatile-oil-yielding plants at present under cultivation and the success of the industry based on these few plants should be sufficient justification for widening the scope of our efforts.

The cultivated plants at the present time are principally the mints, peppermint and spearmint, together with small quantities of such plants as wormwood, tansy, and wormseed.

The distillation of peppermint [a] and spearmint in the United States dates back to 1816. when the peppermint plant was first cultivated for the production of the oil in New York, followed somewhat later by spearmint. The cultivation gradually spread, until at present the center of the industry is in Michigan, with limited production in Indiana.

The cultivation in New York and Michigan has decreased recently, owing to a slight oversupply, which, however, is probably only temporary. Peppermint and spearmint are possibly more largely distilled in the United States than any other oils at the present time, excluding such plants as grow wild and which produce large quantities of oil, notably the turpentine-yielding pines.

The wormwood plant (*Artemisia absinthium*). although introduced from Europe, has been cultivated to some extent commercially in Wisconsin, Michigan, New York, and other North-Central States. The distillation of the oil has been conducted with a certain degree of success, the yield from fresh, flowering herbs being from one-third to one-half of 1 per cent. It is, however, questionable whether, in the light of the recent European agitation against wormwood, this plant will continue to be cultivated for its oil to the same extent as in the past.

The herb tansy (*Tanacetum vulgare*) is grown for its oil in a small way in the eastern part of the United States and yields from one-tenth to one-fifth of 1 per cent of a volatile oil used principally in medicine.

The plant American wormseed (*Chenopodium ambrosioides* L., var. *anthelminticum*) is grown chiefly in Maryland and southward,

---

[a] Bulletin 90, pt. 3, Bureau of Plant Industry, U. S. Dept. of Agriculture.

where the plant is found growing wild. There are produced the seeds, which are valuable commercially, and the volatile oil distilled therefrom, which also possesses the anthelmintic action of the seeds.

Another volatile oil which is produced on a very extensive scale and which has been distilled commercially for more than a century, namely, oil of turpentine, deserves brief mention. The production of turpentine oil is confined principally to the Southern and Gulf States, from Virginia to Florida, regions of extensive pine forests. Turpentine is obtained as an oleoresinous exudation from several varieties of pine trees, chief among which is the long-leaved pine (*Pinus palustris* Miller). Other species, such as *Pinus taeda* L. and *Pinus echinata* Miller, also yield a valuable oleoresin. Unlike most volatile oils, the oil of turpentine is not distilled directly from the plant but results as one of the products of the distillation of the oleoresin obtained from the trees, the other product being the rosin or colophony of commerce. The usefulness and value of oil of turpentine in commerce, both in the arts and in medicine, where it is practically indispensable, require no further comment.

The plants just enumerated represent the principal volatile-oil plants which are cultivated or gathered for oil production in the United States. The distillation of oils from the mint species is a singular instance of an industry of commercial magnitude, while the several other oils which are being distilled from cultivated plants occupy a secondary position in production. The further development of some of the oils mentioned will be controlled largely by the consumption of the products and by the demand which may be created for them.

The experimental work being conducted at the present time at the Arlington Experimental Farm, near Washington, D. C., is such as to demonstrate the practicability of more extensive cultivation of the plants already grown, as well as of other plants growing wild at present, but which by proper methods of domestication can probably be greatly improved both from the standpoint of luxuriance of growth and of fragrance.

The introduction of foreign species of volatile-oil plants and the testing of the same upon native soil are also receiving considerable attention, and the successful production of oil is clearly assured in some cases. Suitable localities, however, must be chosen to conform with the natural habitats of the introduced plants in order to attain the highest degree of efficiency of production.

WILD PLANTS.

Possibly the number of wild aromatic plants which are used in the manufacture of volatile oils exceeds that of those which are at present cultivated. The extent of the production of the oils is much less, chiefly because of the more or less scattered condition of these plants, and therefore the difficulty of gathering them in large quantities. Usually these wild aromatic plants are distributed over wide areas confused largely with other volatile or nonvolatile species, thus causing the rapid collection of the plants to be seriously hindered. For this reason, probably, together with lack of interest in the cultivation of the wild plants, the production of their oils has been largely restricted.

SASSAFRAS.

A specific example of an important uncultivated plant which yields a volatile oil of considerable value is the sassafras tree. Sassafras oil was one of the first volatile oils distilled in America. The range of the tree is from Florida, where it was originally discovered, to Virginia and Pennsylvania, and even as far north as New York and the New England States. It is quite abundant in the South-Central States, especially Kentucky, Tennessee, and Arkansas. The production of this oil attained commercial significance early in the last century, and it is distilled extensively at present in Kentucky, Tennessee, Pennsylvania, Maryland, and Virginia; also to a less extent in Ohio, Indiana, and New York.

Although the distillation of this very fragrant oil, which is obtained principally from the bark of the root of the sassafras tree (*Sassafras officinalis*), has assumed a strong commercial aspect, the tree has not been grown, strictly speaking, for oil purposes. No doubt the great abundance and the ready accessibility of the trees growing wild are the causes of the noncultivation of this tree for commercial purposes. The leaves and branches of the tree are faintly aromatic, but are not used as a source of the oil. The root bark and wood, which contain from 1 to 8 per cent of volatile oil, form the crude source of supply. The oil is distilled by the ordinary method of steam distillation, the wood and bark of the root being previously coarsely comminuted to admit of better extraction.

WINTERGREEN AND SWEET BIRCH.

The distillation of the oils of wintergreen and sweet birch is a further example of wild aromatic plants furnishing oils in sufficient quantities to supply the trade. Both wintergreen (*Gaultheria procumbens*) and sweet birch (*Betula lenta*) occur largely from the New England States and North-Central States to Georgia, Florida, and Alabama. The distillation of these oils dates back nearly as far as that of the oil of sassafras and has developed until the industry at present is of some significance. Wintergreen and sweet birch are entirely unrelated plants, yet the oils produced from them by distillation are for all practical uses identical. Mention has been made previously of the fact that the oil in these plants is formed by reaction and does not preexist in the tissues. The glucosid gaultherin is the constituent which is responsible for the formation of this oil, and since the reaction between this glucosid and the plant ferment is the same in both plants, the resulting volatile oil (or methyl salicylate) must necessarily be similar.

In the case of the sweet birch, which is a tree of some size, the bark of the trunk and the small branches are used for distillation, being previously cut into small pieces and allowed to macerate with water before introduction into the still. A yield of three-tenths to three-fifths of 1 per cent of oil is obtained. On the other hand, for the separation of the oil of wintergreen the leaves and twigs are used, the plant being more or less shrubby. The same treatment is applied to wintergreen as to sweet birch, maceration in water being allowed to continue for a period of several hours prior to distillation. The yield of volatile oil from wintergreen varies from one-half to 1 per cent. Owing to the abundance of these plants their cultivation especially for the volatile oil has not been attempted, the material being collected from the plants as they grow in their native habitats. The strict enforcement of the Food and Drugs Act has tended to curtail largely the use of the synthetic oil (methyl salicylate) for certain purposes where the natural oil is required. A more active demand for the natural oils of sweet birch and wintergreen has necessarily resulted, the price of these oils being thereby materially advanced.

CANADA FLEABANE.

Several other plants capable of yielding volatile oils of some value are at present distilled in the United States. A very common herb growing abundantly in the North-Central and Western States, the Canada fleabane (*Erigeron canadensis*), usually regarded as a weed and known to westerners as the fireweed (not the true fireweed, however), is distilled in a small way in connection with the distillation of peppermint. The plant, which is a hardy annual, is not cultivated, but is cut in the wild condition, no special care being taken to eliminate other aromatic weeds or plants, and consequently there results an oil which, although representing the oil of erigeron, is far below the true standard of the oil, owing to the presence of extraneous plant matter introduced during distillation.

EUCALYPTUS.

The production of eucalyptus oil from the leaves and twigs of the blue-gum tree (*Eucalyptus globulus*) is of considerable importance in the volatile-oil industry of the United States. The commercial production of this oil is confined almost exclusively to the State of California, where the tree grows abundantly. The tree is not cultivated as a source of volatile oil, but is extensively grown for ornamental, fuel, and timber purposes. The leaves and twigs are collected from the waste branches or brush resulting when the trees are cut for timber or wood and used for the purpose of distillation. The material selected for distillation may be coarsely comminuted and the essential oil readily obtained therefrom by the usual method of steam distillation.

The yield of oil varies from three-tenths to four-fifths of 1 per cent, according to the quantity of woody branches and twigs introduced into the still with the leaves, the latter producing the highest yield of oil. The use of this oil is very general, and it is employed chiefly as a therapeutic agent. From 70 to 90 per cent of the oil consists of eucalyptol or cineol, the chief constituent and the one to which its valuable antiseptic properties are due.

The waste leaves and branches accumulating when the trees are cut for lumber or wood are not fully utilized. At points where a considerable number of trees are being felled a distilling apparatus could under favorable circumstances be profitably installed and successfully operated at a very moderate expense. It has been estimated that 2 tons of leaves and twigs will produce from 3 to 4 gallons of oil at a cost of about $3 a gallon for distilling the oil.[a]

MONARDAS.

Two additional plants possessing volatile oils of antiseptic value and growing wild in the whole north-central portion of the United States, from Pennsylvania to Minnesota, are wild bergamot (*Monarda fistulosa*) and horsemint (*Monarda punctata*), belonging to the Labiate tribe. These plants yield oils rich in antiseptic constituents, the former producing an oil consisting chiefly of the liquid phenol carvacrol, while the oil from the latter consists for the most part of the crystalline phenol thymol. Both of these constituents are isomeric in character and of equal value as antiseptics, the extensive use of thymol for medicinal purposes being familiar to most people.

Wild bergamot and horsemint, owing to their hardiness, are capable of profitable cultivation in the North-Central States, where the climatic conditions seem to be especially suitable for their growth and for the production of oil. The whole fresh plant during its flowering condition is generally distilled, the amount of oil obtained being influenced by conditions of growth and culture, but averaging from three-tenths to 1 per cent or more. The perennial nature of the plants enables the grower to produce them from year to year with a mini-

---

[a] Bulletin 196, California Agricultural Experiment Station, p. 34.

mum of labor on somewhat sandy, dry soil which possibly has no great value for the production of other crops.

PENNYROYAL.

Pennyroyal is a small annual herb characteristic of the east-central portion of the United States. It is distilled for its oil principally in Ohio and North Carolina, with smaller operations in intermediate States. The pennyroyal plant (*Hedeoma pulegoides*) is native to the United States, is readily propagated and grown, and yields a volatile oil which finds extensive application in therapeutics. The yield of oil distilled from the fresh flowering herb varies from three-fifths to 1 per cent.

MISCELLANEOUS AROMATIC PLANTS CAPABLE OF CULTIVATION.

The foregoing instances represent typical cases of wild plants indigenous to the United States and capable of yielding volatile oils, some of which are distilled on a quasi-commercial basis while others are not grown or distilled at all.

Hosts of other wild aromatic plants are found growing in all sections of the country, many possessing exceedingly fine fragrance and many, on the other hand, possessing odors less attractive but nevertheless possibly of value. These odorous plants will in most cases produce volatile oils which may contain constituents of value, not only in the perfumery trade but also in the arts and medicine. A systematic canvass of the flora of the United States, with special attention to those plants which possess an aroma, and a trial distillation of the same, followed by a careful, detailed chemical examination of the oils, will no doubt bring to light new oils, the value of which may be determined from the nature of the constituents identified in them. Several new volatile oils have been distilled within the past year which have been shown by chemical analysis to contain highly valuable constituents. The results of these experiments, which have proved very gratifying, will be published in the near future, and the significance of the exploration in this field of research will be clearly indicated. Practically no progress has been made in this direction within the last few decades. The necessity of these investigations is therefore strongly recommended.

Various other plants deserving mention, besides those already cultivated and those growing wild which possess volatile products of value to the perfumer and confectioner, are the rose, lavender, rose geranium, rosemary, thyme, sweet basil, summer savory, and sweet marjoram, and the umbelliferous seeds (caraway, anise, fennel, and coriander), besides the citrus fruits lemon and orange. The plants of the first general class, though not native to this country, have been introduced and grown as garden plants, luxuriant growth and excellent aromas usually being obtained.

The umbelliferous plants mentioned have also been largely grown, although only on a garden scale, usually for their seeds, which possess considerable value to the housewife and to the confectioner for flavoring or condimental purposes. The distillation of the oils from these seeds has been very largely for experimental purposes only.

The citrus fruits, although grown very extensively, have received but slight attention in the United States from the standpoint of their volatile oils, which are of so much value to the scenter and perfumer.

The rose, lavender, and rose geranium, although possessing exceedingly fragrant volatile oils have received only trifling consideration as regards cultivation for the aroma.

It is not unlikely that certain sections of the United States are adapted to the growth of the Bulgarian rose, which produces the rose oil of commerce. In order to locate these desirable regions, practical tests would be required, attention being paid to the quality of the perfume obtained and also to the labor required in the gathering of the rose petals. Besides the usual variety of rose used for perfume cultivation, the *Rosa damascena*, there are a number of other species which have become naturalized in this country and which possess fragrance of exceedingly high quality, besides being prolific bearers.

Experiments in connection with the growing of roses for perfumery purposes are worthy of attention in some of the southern portions of the United States where the conditions of climate are especially favorable and where, since the petals must be plucked by hand for distillation, labor would be sufficiently cheap to insure a certain degree of success.

Lavender (*Lavandula vera*), now grown extensively in the semi-mountainous districts of France and in England for the volatile oil, is no less capable of growth on the soils of this country than other plants which are at present grown profitably. The regions of growth in France, Italy, and England are not entirely dissimilar and do not possess any more suitable climatic and soil conditions than might be supplied in some sections of the United States. In this case experiments would also be necessary to locate desirable regions, but the labor factor would be minimized considerably owing to the fact that the entire tops of the plants are distilled. Owing to the little labor required in connection with lavender, enterprise in this matter should not be lacking.

The rose geranium (*Pelargonium odoratissimum*), a plant with an exquisite odor grown and distilled in France, Spain, Algiers, and the island of Reunion, deserves some consideration with regard to cultivation, inasmuch as the oil distilled from the plant is of such a nature as to make it almost indispensable in the perfumery industry. Unlike that of lavender, the odor of the rose geranium resides in the leaves, the flowers being almost odorless. Experiments in a preliminary way are now being carried on to determine the quality of the oil capable of being distilled from this plant. As in the case of the rose and lavender, the most suitable location can be learned only by a system of tests in localities with different climatic and soil conditions.

Rosemary (*Rosmarinus officinalis*), thyme (*Thymus vulgaris*), sweet basil (*Ocimum basilicum*), summer savory (*Satureja hortensis*), and sweet marjoram (*Origanum marjorana*), besides others of this type originating in Mediterranean countries and yielding oils of excellent fragrance for both the perfumers and the toilet-preparation manufacturers, can by proper attention and perseverance no doubt be produced advantageously. A factor of considerable import in the growth and distillation of these plants is that whole fresh herbs can be distilled, thus obviating the necessity of picking the flowers by hand.

The distillation of oil from such seeds as caraway, anise, fennel, and coriander, which are so universally used for flavoring and scenting purposes, has been successfully exploited in southern Europe for decades. These seeds have been introduced into the United States and grown in small quantities, principally for household use. The ease of production as a household necessity should be sufficient stimulus for growing the plants on a broader basis for the distillation of the very fragrant oils. The North-Central States, with their excellent soil and climate, undoubtedly are capable of producing profitable yields of seeds giving from 2 to 7 per cent of volatile oil. The method of distillation is similar to that of leaves or herbs, with the exception that, in order to facilitate the permeation of the steam, the seeds are ground coarsely before being subjected to the steam vapors.

The commercial isolation of oils from citrus fruits and their by-products centers principally in Sicily and Italy. The production of oil from either lemon or orange peel in the citrus regions of California has received but slight attention and should be deserving of more, inasmuch as the demand for these oils is very constant and the prices reasonably high. The distillation of waste lemons or unsalable lemons would possibly yield a volatile oil of lemon of fair quality, which no doubt would find a ready market. The Sicilian methods of hand expression are practically out of the question because of the labor factor involved. The distillation of lemon-tree prunings yields an oil of extremely high citral content, which should prove valuable for flavoring purposes.

COMMERCIAL ASPECT OF THE INDUSTRY.

VALUE AND CONSUMPTION OF VOLATILE OILS.

Mention has already been made of the value in general of volatile oils as industrial products, which commercially have not been manufactured in the United States to any extent, the mint oils being singular exceptions. Lack of interest in the growth and development of perfumery plants is principally responsible for the inactive condition now existing in this important phase of industrial enterprise. Possibly a lack of experience with regard to the growth of the plants concerned and the methods necessary for success has been largely instrumental in preventing the upbuilding of this branch of industry.

It must be conceded that very large quantities of volatile oils are at present consumed in the United States in the several uses to which they are applied. In the manufacture of perfumes the rôle played by volatile oils is all important. A large proportion of the amounts consumed enters the channels of the perfumery trade. Usually perfumes consist of blends of odors brought about by a skillful combining of several oils in varying proportions through a medium capable of holding in solution these oils and odoriferous ingredients. The manufacture of perfumes has shown but little development in the New World. Perfumery products are largely imported in the prepared condition, chiefly from France, where the skillful art of compounding has been scientifically developed.

The use of volatile oils in flavoring and in the manufacture of flavoring extracts is very extensive, but it is restricted to a comparatively small number of oils, principal among which are lemon, orange, wintergreen, peppermint, and others of this type.

For scenting purposes, such as aromatizing soaps and toilet preparations in general, volatile oils have been employed very extensively in the United States. Their use in this line of application has increased with the increase in the manufacture of these much-demanded articles.

On the other hand, the medicinal value of certain oils and of certain constituents which can be isolated from them has created a demand which in part has been supplied by home production and in part by foreign production. The separation of important therapeutic ingredients, chiefly antiseptics, has been highly serviceable in the treatment of many ailments, a striking instance of this kind being the separation of camphor from the oil of camphor, this ingredient playing an important rôle in medicine as well as in the arts. Other oils deserving mention in this connection are those of eucalyptus and thyme, the former yielding the valuable eucalyptol and the latter thymol. Another example is peppermint oil, from which menthol is isolated.   All of these constituents possess therapeutic value of no little importance.

In order that the grower may become acquainted with the approximate value of volatile oils on the American market, the following tabulation of prices has been prepared.   The perfumery articles listed include the principal volatile oils which enter the markets of the United States for consumption, the prices being current wholesale quotations in effect in January, 1910.   Prices are per pound unless otherwise stated.

Wholesale prices of various volatile oils in the markets of the United States, January, 1910.

Almond, bitter	$3.25 to $4.75
Anise	1.10 to 1.12½
Bay	1.90 to 2.00
Bergamot	3.75 to 4.00
Cade	.16 to .20
Cajeput	.52½ to .55
Camphor	.09 to .10
Caraway seed	1.15 to 1.25
Cedar, leaf	.42½ to .45
Cedar, wood	.16 to .17
Cinnamon	6.50 to 12.00
Citronella	.25 to .28
Cloves	.70 to .72½
Copaiba	1.00 to 1.10
Coriander	5.00 to 6.00
Cubeb	3.00 to 3.25
Erigeron	1.50 to 1.60
Eucalyptus, American	.35 to .60
Fennel seed	1.10 to 1.30
Geranium, rose, African	3.50 to 4.00
Geranium, rose, Turkish	2.25 to 2.50
Ginger	4.00 to 4.50
Ginger grass	1.10 to 1.35
Hemlock	.45 to .50
Juniper, berries	.80 to 1.00
Juniper, wood	.23 to .25
Lavender, flowers	1.85 to 2.25
Lavender, spike	.60 to 1.10
Lemon	.77½ to .85
Lemon grass	.80 to .85
Lime, expressed	1.75 to 2.00
Lime, distilled	.55 to .60
Linaloe	2.80 to 2.85
Mace	.70 to .75
Male fern	1.90 to 2.20
Mustard	3.00 to 4.00
Neroli, petals	50.00 to 75.00
Neroli, bigard	35.00 to 50.00

⁶ Oil, Paint, and Drug Reporter, vol. 77, no. 4, January 24, 1910, p. 32.

Nutmeg	$0.70 to	$0.80
Orange, bitter	2.25 to	2.35
Orange, sweet	2.20 to	2.40
Origanum	.20 to	.40
Patchouli	4.00 to	4.25
Pennyroyal	1.70 to	1.80
Pennyroyal, French	1.40 to	1.50
Peppermint, tins	2.00 to	2.10
Peppermint, bottles	2.30 to	2.35
Petit grain, French	5.00 to	6.00
Petit grain, South American	2.40 to	2.75
Pimento	1.90 to	2.25
Rose, natural_____per oz__	5.00 to	5.50
Rosemary flowers	.67½ to	.75
Safrol		.40
Sandalwood	3.00 to	3.25
Sassafras	.55 to	.65
Savine	1.25 to	1.30
Spearmint	1.75 to	1.85
Spruce	.40 to	.45
Tansy	2.50 to	2.75
Thyme	1.00 to	1.10
Wintergreen (or sweet birch)	1.45 to	1.75
Wintergreen, leaf	3.25 to	4.25
Wormseed	1.50 to	1.60
Wormwood	6.25 to	6.50
Ylang-ylang	47.00 to	65.00

IMPORTS AND EXPORTS OF VOLATILE OILS.

Importations of volatile oils and allied products have increased from year to year until at the present time the expenditures for volatile oils and perfumes aggregate more than $2,000,000 annually.

According to the statistics of imports compiled by the Bureau of Statistics of the Department of Commerce and Labor, the importation of volatile and distilled oils, free and dutiable, for the year ending June 30, 1908, amounted to $3,619,161.33.[a] From this amount there should be deducted $886,923, which represents distilled oils not of plant origin. The total importation, therefore, of volatile oils, free and dutiable, distilled from plants for the above year was valued at $2,732,238.33. These figures represent only the volatile oils imported.

In addition to the sum mentioned, the imports of alcoholic perfumery, including toilet and cologne waters and alcoholic handkerchief perfumes, must be considered. The total imports of this class of perfumes for the year ending June 30, 1908, amounted to $484,498.43.[a]

---

[a] Commerce and Navigation of the United States, 1908, p. 917.

The value of toilet preparations, such as cosmetics, hair washes, dentrifices, pastes, pomades, and powders, into which perfumery substances enter may also be mentioned in this connection. The imports of these preparations for the above year reached a total of $604,258.09.[a]

For purposes of comparison and to illustrate the remarkable increase of consumption of volatile oils of foreign production, the statistics extending over several years are tabulated.[b]

TABLE III.—*Imports of volatile and distilled oils for the years ending June 30, 1903 to 1908, inclusive.*

Free imports from—	1903.	1904.	1905.	1906.	1907.	1908.
Europe	$1,253,360	$1,318,606	$1,387,268	$1,617,796	$2,227,530	$2,215,265
North America	2,747	1,315	16,389	5,713	2,431	5,996
South America	2.364	4,052	2,205	750	4,969	14,886
Asia	191,730	252,729	176,563	808,781	407,008	314,688
Oceania	129					
Africa		290	24		304	
	1,450,380	1,576,992	1,582,449	1,983,040	2,642,242	2,550,835

TABLE IV.—*Imports of volatile and distilled oils for the years ending June 30, 1903 to 1908, inclusive.*

Dutiable imports from—	1903.	1904.	1905.	1906.	1907.	1908.
Europe	$590,493	$745,013	$865,008	$850,989	$987,919	$1,028,630
North America	14,444	12,210	4,994	12,794	18,879	15,678
South America				15		415
Asia	86,768	41,214	54,296	38,361	32,572	22,441
Oceania	14,296	20,958	24,343	15,529	17,123	19,308
Africa		361	3,003	12,227	3,485	8,134
Total	706,001	819,756	951,644	929,915	1,059,978	1,094,606

The steady increase in the importation of perfumery products, as shown in Tables III and IV, indicates that the consumption of volatile oils and scenting materials in America is also increasing. With the exception of peppermint, comparatively small quantities of crude oils are distilled and exported from the United States. The exports of peppermint oil, distilled largely in New York and Michigan, for the year ending June 30, 1908, were 141,617 pounds, valued at $357,555,[c] while all other essential oils exported amounted to $214,765.

The imports of volatile oils and perfumery materials far exceed the exports of the same products, the principal product of export being peppermint oil, a singular case where the distillation approaches industrial size in the United States.

---

[a] Commerce and Navigation of the United States, 1908, p. 919.

[b] Commerce and Navigation of the United States, 1908, p. 279.

[c] Commerce and Navigation of the United States, 1908, p. 636.

The total yearly outlay for the crude materials, and also for the finished products, is sufficient to attract attention and is deserving of concerted action on the part of growers and others who might profitably engage in this neglected field of research and practice.

The present source of these commercial products, which may be gleaned from the tabulation, is Europe, from whence they are imported both in the crude state and in the manufactured condition. Italy possibly furnishes the smallest quota of volatile oils and the largest valuation, the products being chiefly the citrus oils, supplied solely by Sicily and Italy and consumed to a great extent in the United States. From France the large proportion of perfumery extracts and finer essential oils is imported. while Germany, Turkey, and Great Britain distribute to this country large consignments of crude and purified volatile oils.

The Mediterranean regions of Europe are the chief sources of these aromatics, which are so generally employed in the industries in diverse ways. The cost of production is minimized in these countries because of the cheaper class of labor as compared with labor in America, for instance. In the handling of many flowers and plants, much hand labor is required. especially in the collection of the material prior to distillation. The actual distillation and purification of the oils can be conducted with equal economy in the United States, while in the case of no small number of plants which may be suitably collected and distilled in the whole condition the question of labor becomes a less serious factor, especially in some instances where mowing machines may be employed advantageously to harvest the crops. Where hand picking is required, as in the case of some of the more delicate odors from flowers and flowering tops, cultivation and extraction of the odor could possibly be carried out in the Southern States, which have abundant sunshine, an important prerequisite in odor development. Furthermore, the labor conditions in the Southern States are such that the cost of gathering, which is a serious obstacle, would be comparable to a degree with that in foreign countries

CONCLUSIONS.

In view of the success which has been achieved in the United States along a number of special lines, the outlook for a very considerable extension of the volatile-oil industry in general seems promising. Favorable conditions of soil and climate seem to be obtainable. With an increased practical knowledge of how to handle the crops of greatest promise and with a working familiarity with the forms of apparatus used in separating the oils, the preliminary steps leading to such an extension will have been taken. Before a full-fledged industry can be expected to appear, however, much preliminary experimental work must be done over a wide area in order to ascertain the most successful combinations of soil, climate, and labor conditions.

From the standpoint of the consumption of products derived from volatile oils obtained from plants, the commercial statistics show a large and active market. They also show that the demand is now supplied in very large part from foreign sources, and an active interest in testing the possibilities of our land is suggested.

INDEX.

O

www.ingramcontent.com/pod-product-compliance
Lightning Source LLC
Chambersburg PA
CBHW021544270326
41930CB00008B/1361